Arduino
A Beginner's Guide to Programming Electronics

By Chuck Hellebuyck

**Published by Chuck Hellebuyck's Electronic Products.
Copyright 2016, All rights reserved.**

No part of this publication may be reproduced, stored in a retrieval system, or transmitted in any form or by any means, electronic, mechanical, photocopying, recording, or otherwise, without the prior written permission of the publisher.

The Arduino name and logo are registered trademark of Arduino LLC.

The schematic pictures were created with software from Fritzing.org.

All other trademarks mentioned herein are the property of their respective companies.

Printed in the United States of America.

Table of Contents

Chapter 1 – What is Arduino 7
    Arduino Uno Overview 8
    How to Get Arduino Running on Mac 17
    How to Get Arduino Running on Windows 21
    Arduino Development Environment 21
    Arduino C Compiler 25
    Getting Started 33

Chapter 2 – Flash an External LED 35
    Hardware 35
    Software 37
    How It Works 37
    Next Steps 40

Chapter 3 – Train Crossing 41
    Hardware 41
    Software 43
    How It Works 43
    Next Steps 45

Chapter 4 – LED Traffic Light 47
    Hardware 47
    Software 49
    How It Works 50
    Next Steps 52

Chapter 5 – Scroll LEDs 53
    Hardware 53
    Software 55
    How It Works 56
    Next Steps 58

Chapter 6 – Sensing a Switch 59
    Hardware 60
    Software 61
    How It Works 62
    Next Steps 65

Chapter 7 - Read a Potentiometer 67
Hardware 67
Software 69
How It Works 70
Next Steps 73

Chapter 8 - Sensing Light 75
Hardware 75
Software 77
How It Works 77
Next Steps 79

Chapter 9 – Creating Sound 81
Hardware 81
Software 83
How It Works 83
Next Steps 85

Chapter 10 – Dimming a LED with PWM 87
Hardware 87
Software 89
How It Works 89
Next Steps 91

Chapter 11 – Serial Communication 93
Hardware 93
Software 95
How It Works 96
Next Steps 98

Chapter 12 – Liquid Crystal Display 99
Hardware 99
Software 101
How It Works 102
Next Steps 104

Chapter 13 - Arduino Libraries 107
Custom 3rd Party Libraries 108
Manual Library Installation 111

Conclusion 113

Appendix A – Parts List for Projects 115
Appendix B – CHIPINO Demo Shield 117
Appendix C – DFRobot LCD Shield 119

Chapter 1 – What is Arduino

You can purchase many different microcontroller based modules for electronic development but the easiest C programmable module is the Arduino. Many of the competitors rely on a BASIC language compiler or interpreter to handle the software duties but the Arduino goes a step further by using a C compiler with many pre-written functions to make it easier for the beginner to use. This accomplishes two tasks; it allows complete beginners to get started and it also teaches them the fundamentals of the C language that is used in industry.

The Arduino in many ways was developed for the artist or mechanical hobbyist crowds who aren't your normal electronics users. It's proven by the fact that software written for the Arduino is called a Sketch (like an artist's drawing) rather than a program, which is the typical name used for software. The Arduino platform is all open-source so all the electrical schematics, circuit board layouts and software are all free to download from the website: http://arduino.cc. There are various versions of the Arduino board and the latest design is the Arduino Uno. In fact all the projects in this book will use the Uno version. To better understand the Arduino let me detail what makes up the Arduino Uno module.

Arduino Uno Overview

The Arduino Uno is a microcontroller board based on the Atmel ATmega328P microcontroller. It has 14 digital input/output pins (of which 6 can be used as PWM outputs), 6 analog inputs, a 16 MHz crystal oscillator, a USB connection, a power jack, an In Circuit Programming (ICP) header, and a reset button. It contains everything needed to support the microcontroller so all you have to do is connect it to a computer with a USB cable to get started. The Uno can be powered from the USB cable or separately from an AC-to-DC adapter or battery. To program the board you need the Arduino programming software, which I'll explain in a few pages. The Uno is shown in Figure 1-1.

Figure 1-1: Arduino Uno

Module Details

The schematic for the board and the board layout can be downloaded from the arduino.cc website under the products section. Just click on the UNO picture.

The features of the module are summarized below:

Microcontroller	ATmega328
Operating Voltage	5V
Input_Voltage (recommended)	7-12V
Input Voltage (limits)	6-20V
Digital I/O Pins	14 (of which 6 provide PWM output)
Analog Input Pins	6
DC Current per I/O Pin	40 mA
DC Current for 3.3V Pin	50 mA
Flash Memory	32 KB (ATmega328) of which 0.5 KB used by bootloader
SRAM	2 KB (ATmega328)
EEPROM	1 KB (ATmega328)
Clock Speed	16 MHz

Power

The Arduino Uno can be powered via the USB connection or with an external power supply. The power source is selected automatically by sensing circuitry within the Uno. The external power can come either from a 7-12 volt AC-to-DC adapter with a 2.1mm center-positive plug or a 9 volt battery with a cable that ends in a 2.1mm plug. Many of the Arduino resellers offer this cable or you can easily build one. You can also connect the battery leads to the Gnd and Vin pin header sockets of the POWER connector shown on the lower left side of Figure 1-1.

The POWER connector has the following pins:

VIN – This pin his connected to the center tap of the 2.1 mm connector and makes it easy to connect to the input voltage of the Arduino board.
5V - This pin has the regulated 5 volt power produced by the Arduino's 5 volt regulator.
3V3 - A 3.3 volt supply is generated by the on-board 3.3v regulator. Maximum current draw is 50 mA.
GND - Ground pins.

Memory

There are three areas of memory in any Arduino board;
1) Flash memory (program space), where the Arduino sketch is stored.
2) Static RAM (SRAM) is where variables are stored during power on. RAM is typically erased or scrambled after power is removed.
3) EEPROM memory where data or variable data can be stored and not lost after power is shut off.

Flash Memory: 32 KB of which 0.5 KB used by bootloader
SRAM: 2 KB
EEPROM: 1 KB

Input and Output

Each of the 14 digital pins on the Uno can be used as inputs or outputs, using pinMode(), digitalWrite(), and digitalRead() functions. Each pin can provide or receive a maximum of 40 mA and has an optional internal pull-up resistor (disconnected by default) of 20-50 k ohms. In addition, some pins have specialized functions:

Serial: 0 (RX) and 1 (TX) – These pins are used to receive (RX) and transmit (TX) serial data and are connected to the corresponding pins of the FTDI USB-to-TTL Serial chip. It's through these pins that the sketch is loaded into program memory.

External Interrupts: 2 and 3 - These pins can react to an external signal automatically if setup in software to do so. They can be configured to trigger an interrupt on a low value, a rising or falling edge, or a change in value. The attachInterrupt() function is used to set these up.

PWM: 3, 5, 6, 9, 10, and 11 - Provide 8-bit PWM output with the analogWrite() function.

SPI: 10 (SS), 11 (MOSI), 12 (MISO), 13 (SCK) - These pins support SPI communication though are not currently included in the Arduino language.

LED: 13 - There is a built-in LED connected to digital pin 13. When the pin is HIGH the LED is on, when the pin is LOW it's off.

Analog to Digital Conversion (ADC): 0 thru 5 - The Uno has 6 analog inputs with 10 bit resolution (i.e. 1024 different values). By default they measure from ground to 5 volts, though is it possible to change the upper end of their range using the AREF pin and the analogReference() function.

I^2C: 4 (SDA) and 5 (SCL) - Support I^2C communication using the Wire library.

Analog Reference: AREF - Reference voltage pin for the analog inputs. Used with analogReference().

Micro Reset: RESET – This pin can be set LOW to reset the microcontroller. The reset switch will pull this pin low when pressed.

Port Registers

These are locations inside the microcontroller that control the state of the I/O pins. Each port register is 8 bits wide. Each port is actually controlled by three separate registers, which are also defined as variables in the Arduino language.

The Data Direction Register (DDRx) determines whether the pin is an INPUT or OUTPUT and controlled by the pinMode() function.

The PORTx register controls the pins that are set to output and can be a 1(HIGH) or 0 (LOW) using the digitalWrite() function.

The PINx register reads the state of pins set as input using the pinMode() function.

The Arduino ports and registered are defined below.

PORTD maps to Arduino digital pins 0 to 7
DDRD - The Port D Data Direction Register - read/write
PORTD - The Port D Data Register - read/write
PIND - The Port D Input Pins Register - read only

PORTB maps to Arduino digital pins 8 to 13
The two bits (6 & 7) map to the crystal pins and are not usable.
DDRB - The Port B Data Direction Register - read/write
PORTB - The Port B Data Register - read/write

PINB - The Port B Input Pins Register - read only

PORTC maps to Arduino analog pins 0 to 5.
DDRC - The Port C Data Direction Register - read/write
PORTC - The Port C Data Register - read/write
PINC - The Port C Input Pins Register - read only

Each bit of these registers corresponds to a single pin; e.g. the low bit of DDRB, PORTB, and PINB refers to pin PB0 (digital pin 8).

Communication

The Arduino Uno has a number of options for communicating with a computer, another Arduino, or other microcontrollers. The Arduino chip has an internal UART TTL (5V) serial communication peripheral which is available on digital pins 0 (RX) and 1 (TX). A USB to Serial chip on the board converts this serial communication to USB. The drivers (included with the Arduino software) provide a virtual com port to software on the computer.

The Arduino software also includes a serial monitor, which allows simple ASCII data to be sent to and from the Arduino board. The USB to Serial chip has RX and TX LEDs on the board that flash when data is being transmitted via the FTDI chip.

A SoftwareSerial library is also available that offers software functions for serial communication on any of the Uno's digital pins.

The Arduino software also supports I2C communication using the Wire library functions. SPI communication is hadled by the SPI library.

Programming

The Arduino Uno control code is written using the Arduino programming software running on a PC, MAC or Linux, which is also called the Arduino Environment. The Arduino chip is programmed using a bootloader or software method so you don't need a separate hardware programmer. The Arduino software has an editor for writing the programs and then a built in compiler and bootloader interface. With a single click of the mouse, your software will be compiled into 1's and 0's and then sent to the program memory of the Arduino chip's program memory.

For advanced users you can also bypass the bootloader and program the microcontroller through the ISP (In-Circuit Serial Programming) header with an extra hardware programmer.

USB Overcurrent Protection

The Arduino Uno has a resettable polyfuse that protects your computer's USB ports from shorts and overcurrent. Although most computers provide their own internal protection, the fuse provides an extra layer of protection. If more than 500 mA is applied to the USB port, the port will shut down or the fuse will automatically break the connection on the UNO until the short or overload is removed. The best solution is to supply external power to the Arduino board rather than rely on the computer motherboard to supply power. Relying on the USB port for power is convenient but an external source such as a simple 9-volt battery or power adapter plugged into the power input is the safer route. Many Arduino sellers offer 9-volt battery cables that will plug right into the power port.

Physical Characteristics

The Uno PCB is 2.7 and 2.1 inches respectively, with the USB connector and power jack extending a bit beyond these dimensions. Four holes allow the board to be attached to a surface or case. If you want to make your own stack on board (known as a shield in the Arduino world) recognize that the distance between digital pins 7 and 8 is 0.16" while all the other pins are 0.100" spacing. This little offset in those pins makes it a little more difficult to create a shield board from an off the shelf protoboard from Radio Shack or similar.

How to Get Arduino Running on Mac

The arduino.cc website has a great step by step guide for installing Arduino on a MAC under the Getting Started menu but I'll cover the basics here. These are the fundamental steps to get Arduino running on a MAC:

- Download the Arduino environment
- Install the Software
- Connect the board
- Run the Arduino environment
- Create a Blink LED project
- Run the Blink project on a Arduino

1 | Download the Arduino environment

The software can be downloaded by clicking on the MAC OS X link on the software download page at Arduino.cc.

2 | Install the Software

Drag the Arduino application icon into the Applications folder. All the required drivers will be automatically installed.

3 | Connect the board

Connect the USB cable to one of the MACs USB ports and then to the USB connector of your Arduino board. You also need to connect power if you are using an external source such as a 9-volt battery. The easiest choice is to power from USB, but to protect your computer I recommend the external source.

4 | Run the Arduino environment

Double-click the Arduino icon to launch the Arduino environment and you should see a screen similar to the one in Figure 1-2.

5 | Create a Blink LED project

Open the LED blink example sketch at File>Examples>1.Basics>Blink. This will open this simple test sketch to flash the LED on the UNO board.

Figure 1-2: Arduino Environment

6 | Run the Blink project on Arduino

Click the "Upload" button in the environment, which is the sideways arrow shown highlighted in Figure 1-3. Wait a few seconds and you should see the RX and TX LEDs on the board flashing. If the upload is successful, the message "Done uploading." will appear in the status bar.

Figure 1-3: Upload sketch to Arduino board

A few seconds after the upload finishes, you should see the pin 13 LED on the board start to blink. If it does, congratulations! You've gotten Arduino up-and-running. If not, go back through the steps to see if you missed something.

The Arduino has an LED already wired to pin 13 so you didn't need to connect any circuitry to the Arduino board. In future projects you will need to connect the proper components to get the sketch to work.

How to Get Arduino Running on Windows

The arduino.cc website has a great step-by-step guide for installing Arduino on a PC under the Getting Started menu. It's the same as the MAC version except you have to install the USB driver. I find the MAC much easier for the beginner to use and it's what I used in this book for my projects. Here are the steps to install it on a Windows PC.

- Download the Arduino environment
- Install the USB drivers
- Connect the board
- Run the Arduino environment
- Create a Blink LED project
- Run the Blink project on a Arduino

Arduino Development Environment

The Arduino development environment contains a text editor where you for write your Arduino sketch. A toolbar with buttons for common functions is at the top of the screen along with a series of menus. There is a message area at the bottom to give feedback when saving or exporting a file. It also displays any code errors. The toolbar functions are explained here.

Verify
Checks your code for errors.

Upload
Loads the sketch onto the UNO

New
Creates a new sketch.

Open
Presents a menu of all the sketches in your sketchbook.

Save
Saves your sketch.

Serial Monitor
The ICON to the far right of all the others opens the Serial Monitor window.

Sketchbook

The Arduino environment uses the term sketchbook to contain all the files in a project. The first time you run the Arduino software, it will automatically create a directory for your sketchbook. You can view or change the location of the sketchbook location from with the preferences setting within the Main menu.

Tabs, Multiple Files, and Compilation

The environment allows you to break up your sketch into multiple files that get combined into one big file when you are ready to program the Arduino hardware. The environment has tabs so each file can appears in its own tab. These can be normal Arduino code files (no extension), C files (.c extension), C++ files (.cpp), or header files (.h).

Uploading

As described in the blink example earlier, to upload your sketch, you need to select the correct items from the **Tools > Board** and **Tools > Serial Port** menus. Once you've selected the correct serial port and board, press the upload button in the toolbar or select the **Upload** item from the menu.

Libraries

Libraries are the heart of what makes the Arduino easy to use with the hardware. These prewritten sections of code you could be included in your sketch for added functionality. To use a library in a sketch, select it from the **Sketch > Include Library** menu. This will insert one or more **#include** statements at the top of the sketch and compile the library with your sketch. Because libraries are uploaded to the board with your sketch, they increase the amount of space it takes up. If a sketch no longer needs a library, simply delete its **#include** statements from the top of your code.

Some libraries are included with the Arduino software while others can be downloaded from a variety of sources. Many people offer libraries they have written themselves. To install one of these most common third-party libraries, use the **Sketch > Include Library > Manage Libraries** to see the list of libraries you can add. You can also search for libraries in the list using the search window at the top of the screen. Chapter 13 offers more details about adding 3rd party libraries.

Serial Monitor

The Arduino can send back serial data through the same programming connections of the USB cable. The data sent can be displayed on the Serial Monitor built into the Arduino Environment. To use the Serial Monitor you need to choose the baud rate from the drop-down that matches the rate passed to **Serial.begin** in your sketch. To send data to the board, enter text and click on the "send" button or press enter.

Preferences

Some preferences for how the IDE will operate can be set in the preferences dialog (found under the **Arduino** menu on the Mac, or **File** on Windows and Linux).

Arduino C Compiler

The Arduino uses a programming language based on C but adds simple built-in functions that are not typical to C programming. These pre-built functions make it easier for the beginner to get started with programming the Arduino. The Arduino sketch gets converted into the 1's and 0's that the Arduino chip needs. This is done by a compiler built into the IDE.

Sketch

A sketch is the name that Arduino uses for a program. It's the software that is written, loaded and run on an Arduino board. The software below is a sample sketch for flashing the built-in LED conneted to pin 13 of the module. I'll describe the various sections as we go through this chapter.

```
/*
Blink
Turns on an LED on for one second, then off for one second,
repeatedly.
The circuit:
LED connected from digital pin 13 to ground.

Note: On most Arduino boards, there is already an LED on the
board connected to pin 13, so you don't need any extra
components for this example.
*/

int ledPin = 13;   // LED connected to digital pin 13

// The setup() runs once, when the sketch starts
void setup()
    {
        pinMode(ledPin, OUTPUT); //initialize the digital pin as
                                 // an output
    }

// The main loop runs over and over again until power is removed
void loop()
    {
        digitalWrite(ledPin, HIGH);   // set the LED on
        delay(1000);                  // wait for a second
        digitalWrite(ledPin, LOW);    // set the LED off
        delay(1000);                  // wait for a second
    }
```

Comments

The first few lines of the blink sketch are a comment:

```
/*
Blink
Turns on an LED on for one second, then off for one second,
repeatedly.
The circuit:
LED connected from digital pin 13 to ground.

Note: On most Arduino boards, there is already an LED on the
board connected to pin 13, so you don't need any extra
components for this example.
*/
```

Everything between the /* and */ is considered a comment block and does not get compiled for the Arduino when it runs the sketch. Comment blocks are used to describe what the program does, how it works, or why it's written the way it is. It's a good practice to comment your sketches, and to keep the comments up-to-date when you modify the code.

There's another style for comments. They start with // and continue only until the end of the line.

```
int ledPin = 13;         // LED connected to digital pin 13
```

The section "LED connected to digital pin 13" is a treated as a comment and not compiled into the Arduino.

Sketch Variables

A variable is a place for storing a piece of data. It has a name, a type, and a value. For example, the line from the Blink sketch declares a variable with the name "ledPin".

It is created as a type "int" or integer and initialized with the value of 13. It's used to indicate which Arduino pin the LED is connected to.

Every time the name ledPin appears in the code, its value will be retrieved. In this case, the person writing the program could have chosen not to bother creating the ledPin variable and instead have simply written 13 everywhere but then later if they wanted to change the pin connection they would have to replace it multiple times. Using the variable "ledpin" allows them to change it at only one place. This is an example of using the variable as a constant.

Often, however, the value of a variable will change while the sketch runs. For example, you could store the value read from an input into a variable. There's more information on variables later on.

setup() and loop()

There are two special functions that are a part of every Arduino sketch: setup() and loop(). The setup() is called once, when the sketch starts. It's a good place to do setup tasks like setting pin modes or initializing libraries. The loop() function is called over and over and is heart of most sketches. You need to include both functions in your sketch, even if you don't need them for anything.

The setup() section in the LED example is shown below. Within the curly brackets is where the ledPin is setup to be an output pin to drive the LED. This only needs to be done once so it's in the setup () section.

```
// The setup() runs once, when the sketch starts
void setup()
    {
    pinMode(ledPin, OUTPUT); //initialize the digital pin as
                             // an output
    }
```

The loop() section in the LED example is below. The digitalWrite and delay(*milliseconds*) pre-built functions are used to control the LED and how fast it blinks. The LED on the ledPIN is drive HIGH or on and then a 1000 millisecond delay (or 1 second) is performed and then the ledPin is driven LOW. This shuts off the LED and another 1 second delay occurs. This routine loops over and over continuously as long as the Arduino has power.

```
// The main loop runs over and over again until power is removed
void loop()
    {
    digitalWrite(ledPin, HIGH);   // set the LED on
    delay(1000);                  // wait for a second
    digitalWrite(ledPin, LOW);    // set the LED off
    delay(1000);                  // wait for a second
    }
```

{ } Curly Braces

Curly braces or curly brackets are a major part of the C programming language that the Arduino software is based on. An opening curly brace "{" must always be followed by a closing curly brace "}" or you will get an error when you compile. Beginning programmers, and programmers coming to C from the BASIC language often find using braces confusing. After all, the curly braces replace the RETURN statement in a subroutine, the ENDIF statement in a conditional and the NEXT statement in a FOR loop. So think of curly braces as the beginning and end of any function. Here are some main uses of curly braces or brackets.

Functions
```
 void myfunction(datatype argument)
{
   statements(s)
 }
```

Loops
```
 while (boolean expression)
 {
   statement(s)
 }

 do
 {
   statement(s)
 } while (boolean expression);
```

```
for (initialisation; termination condition; incrementing expr)
{
   statement(s)
}
```

Conditional statements

```
if (boolean expression)
{
   statement(s)
}

else if (boolean expression)
{
   statement(s)
}
else
{
   statement(s)
}
```

; semicolon

A command line inside your sketch is called a statement in the C language. Each statement needs to mark the end with a semicolon. A statement line can extend multiple lines in your sketch and end with a semi-colon to let the compiler know where to end. The most common error a beginner will see is a missing semi-colon. Two statements will be combined by the compiler and you'll get an unrecognized statement error because of that missing semi-colon.

A simple example of how to end a statement with a semicolon.

int a = 13;

#define

#define is a useful way to create a nickname to a constant value before the program is compiled. Defined constants in Arduino don't take up any program memory space on the chip. The compiler will replace any references to these nicknames with the associated constant value at compile time. This can have some unwanted side effects though, if for example, a constant name that had been #defined is found in some other section of the program and is replaced by the #defined nickname. To prevent this Arduino also offers the const keyword and is the preferred method for defining constants.

```
Format:
```

```
#define constantName value
```

Example

#define ledPin 3
// The compiler will replace any mention of ledPin with the value 3 at compile time.

Note: #define is one of the few times a semi-colon is not required. This is because it is processed before the compiler is launched.

#include

#include is used to insert external files that contain definitions of pre-built functions. This is also known as including a library. One of the advantages to the C language is the ability to easily share functions and grouping common functions together is known as creating a library of functions or just a library. The pre-built functions require a prototype definition which tells the compiler what the function format is.

The list of prototypes for all the pre-built functions included in a library is typically listed in a header file or .h file. The #include directive is used to include that header file with your other sketch files. The #include allows you to include both Arduino libraries and standard C language libraries. The #include in exercised before the compiler is run so the #include line does not end with a semi-colon.

Getting Started

I probably went too deep into programming for the beginner but you'll have a place to reference as you get more qualified at using the Arduino. It's time to start actually programming the UNO and I'll show you several examples of how to make the Arduino work. Let's go back and re-run the blink program with a different pin and an external LED.

All the projects in this book can be built with very common components but if you don't want to build the hardware and just want to program then the CHIPINO Demo-Shield contains all the components pre-soldered in place so you can plug it into the Arduino and just start programming. More details about the Demo-Shield are shown in Appendix B at the back of the book.

CHIPINO Demo Shield

Chapter 2 – Flash an External LED

The first project every beginner needs to do is get an LED to flash. This proves out many things: 1) The hardware is connected properly 2) The software compiles without errors 3) The module can communicate and be programmed from the computer thru a USB connection. 4) We are ready to move on to more complex projects.

If at any point you can't get a project to run on the Arduino, flashing an LED is always a good indicator that the basic setup is working properly.

Hardware

Connect the Arduino UNO per the drawing in Figure 2-1. The long lead of the LED is connected to pin 12, the short lead is connected to the 220 ohm resistor. The other end of the resistor is connected to the Gnd pin of the Arduino UNO

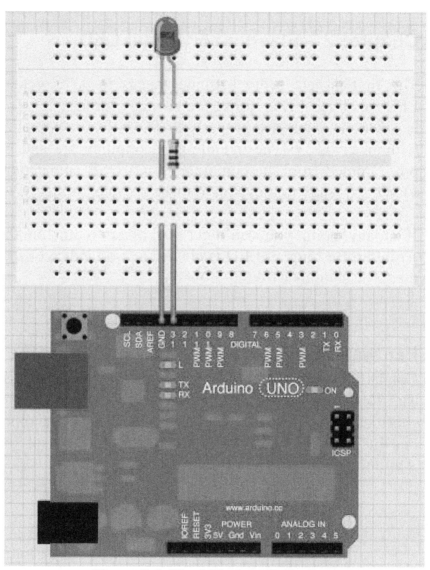

Figure 2-1: Final Flash LED Project

Software

```
/*
  Blink
  Turns on an LED on for one second,
  then off for one second, repeatedly.

*/

void setup() {
  // initialize the digital pin 12 as an output.

  pinMode(12, OUTPUT);
}
void loop() {
  digitalWrite(12, HIGH);    // set the LED on
  delay(1000);               // wait for a second
  digitalWrite(12, LOW);     // set the LED off
  delay(1000);               // wait for a second
}
```

How It Works

The software is quite simple but still requires some explanation. The top of the software contains a header block that describes what the sketch will do. Any time you want to add a block of text that goes over many lines, just place the text between the /* and */ characters. This indicates to the compiler that the text is just comments and not code.

```
/*
  Blink
  Turns on an LED on for one second,
  then off for one second, repeatedly.

*/
```

The first code section of any Arduino sketch is the setup section where any commands are placed that need to run only once before running the main loop. Typically this will be setting up the digital I/O pins and presetting any variable values. Everything between the curly brackets are part of the setup routine.

```
void setup() {

}
```

Within the setup section we can place comments that are ignored by the compiler but allow us to make notes for future reference. A double backslash before the comment is all you need to add to make any text a comment. You have to keep it in one line though. If the text wraps around to a second line then the /* and */ should be used.

The digital pins need to be setup to either be an input or an output. This is done with the pinMode function. The parameters in the parenthesis select the pin and the direction. We want to control the LED with a high or low output signal from pin 12 so we set pin 12 to an output using the line below.

```
pinMode(12, OUTPUT);
```

The main sketch loop is the section of code that runs over and over again. All the command lines need to be contained within the curly brackets. The main loop will run continuously as long as power is not removed from the Arduino or the reset switch is pressed.

```
void loop() {

}
```

The first control is the digitalWrite function that drives pin 12 high which lights the LED.

```
digitalWrite(12, HIGH);    // set the LED on
```

The next line is the delay function. This just creates a one second delay as the value 1000 represents 1000 milliseconds or one second.

```
delay(1000);               // wait for a second
```

The sketch then turns the LED off by setting the same pin low.

```
digitalWrite(12, LOW);     // set the LED off
```

An additional delay function line delays another second.

```
delay(1000);               // wait for a second
```

The sketch then jumps back to the top of the loop since the second curly bracket is encountered. This operation will repeat over and over again to create a simple blinking light.

Next Steps

Simple next steps are to change the pause value to a lower number to flash the LED faster. You could also connect the LED to a different pin and then change the number in the high and low command lines to make that new connection pin flash the LED. You could also control two LEDs at once with a second digitalWrite line. In fact the next project does that to create a train crossing light.

Chapter 3 – Train Crossing

Flashing an LED is a great place to start but let's put it to some practical use. Have you ever stopped at a train crossing and seen the red lights flash back and forth to indicate a train is coming? This can be accomplished real easy with two LEDs. This project can also be built into a model train railroad crossing sign to make it a little more realistic.

Hardware

The hardware is similar to the blink LED project except now there are two LEDs to control. The LEDs are driven on by a high signal on the digital pin that drives them. A low shuts them off. A resistor is placed in series with the LEDs to limit the current. A 1k ohm resistor works fine.

The left LED is connected to the digital pin number 13. The right LED is connected to pin 10. A series resistor is used to limit the current and any value from 220 ohms to 1k ohms can be used. Both LEDs are red on the CHIPINO demo shield.

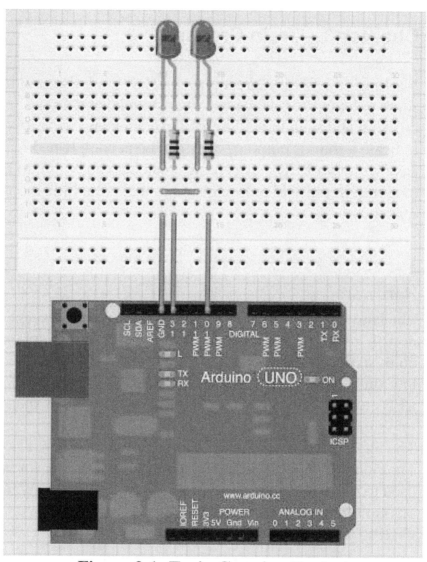

Figure 3-1: Train Crossing Project

Software

```
/*
 Train Crossing
 Blinks two LEDs back and forth like a train
crossing
 */

void setup() {
  pinMode(13, OUTPUT);
  pinMode(10, OUTPUT);
}

void loop() {
  digitalWrite(13, HIGH);// set the left LED on
  digitalWrite(10, LOW); // set the right LED off
  delay(1000);           // wait for a second
  digitalWrite(13, LOW); // set the LED off
  digitalWrite(10, HIGH);// set the right LED on
  delay(1000);           // wait for a second
}
```

How It Works

The first section is just a comment header to describe the sketch.

```
/*
 Train Crossing
 Blinks two LEDs back and forth like a train
crossing
 */
```

The setup loop establishes the two pins to be used as outputs.

```
void setup() {
  pinMode(13, OUTPUT);
  pinMode(10, OUTPUT);
}
```

The main loop controls the two separate LEDs by controlling the digital pins they are connected to. Setting pin 13 to high and pin 10 to low lights the left LED. A delay of 1000 keeps the LED on for 1 second.

```
void loop() {
  digitalWrite(13, HIGH);// set the left LED on
  digitalWrite(10, LOW);  // set the right LED off
  delay(1000);            // wait for a second
```

The second section does the opposite and lights the right LED on pin 10 and shuts off the right LED on pin 13. Another delay of 1000 is added.

```
  digitalWrite(13, LOW);  // set the LED off
  digitalWrite(10, HIGH);// set the right LED on
  delay(1000);            // wait for a second
}
```

The program stays in a continuous loop alternating the lighting of the LEDs to create the train crossing display.

Next Steps

The delay can be changed to make the LEDs flash faster or slower. As mentioned these could be built into a model train crossing sign to make it realistic looking.

Chapter 4 – LED Traffic Light

A traffic light is something we see daily but with the Arduino UNO we can build our own with a few LEDs and resistors. The trick is to have the UNO control each different color LED with a separate digital pin. This way the software can control which LED is on and how long it stays on. Once we have it built then we can control the operation with software.

Hardware

The connections to the LED are similar to the connections in the flash LED example. The long lead of the LED is the positive or anode pin. This is connected to the UNO digital pin that will control it. The green LED is connected to pin 12, the yellow LED to pin 11 and the red LED to pin 10. A resistor is in series between the digital pin and the LED anode lead. This is needed to limit the current so the LED doesn't burn-up.

The shorter lead of the LED is the cathode or negative lead. All the LEDs cathodes are connected together through jumper wires. They connect to the ground pin on the same digital pin header. When the digital pin connected to an LED is high, that LED will light. When a digital pin is low, the LED is off.

The hardware connections are shown in Figure 3-1. Red is on the right, yellow in the middle and green on the left.

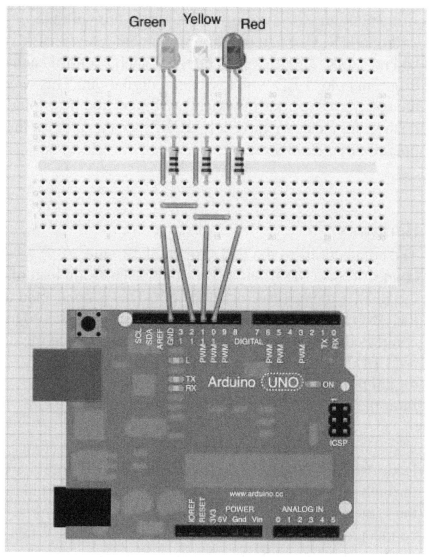

Figure 4-1: Traffic Light Project

Software

```
/* Traffic Light using Red, Green and Yellow LEDs
all controlled by digital pins on the UNO */

const int Red = 10;
const int Yellow = 11;
const int Green = 12;

void setup() {
 pinMode(Red, OUTPUT);
 pinMode(Yellow, OUTPUT);
 pinMode(Green, OUTPUT);

}

void loop() {
   digitalWrite(Red, HIGH);      // set the LED on
   digitalWrite(Yellow, LOW);    // set the LED off
   digitalWrite(Green, LOW);     // set the LED off
   delay(2000);            // wait for two second

   digitalWrite(Red, LOW);       // set the LED off
   digitalWrite(Yellow, LOW);    // set the LED off
   digitalWrite(Green, HIGH);    // set the LED on
   delay(2000);            // wait for two second

   digitalWrite(Red, LOW);       // set the LED off
   digitalWrite(Yellow, HIGH);   // set the LED on
   digitalWrite(Green, LOW);     // set the LED off
   delay(500);    // Wait for half second
}
```

How It Works

The first section is actually before the setup section. This code snippet creates labels for the digital pins by using the "const" directive which is short for constant and the "int" directive which is short for integer. The label "Red" is associated with the digital pin number 10. It cannot change while running the sketch since the "const" makes the value constant. Anytime the label Red is used in the Sketch, the compiler with automatically replace it with the number 10 before it produces the code that gets sent to the UNO.

By creating a label for each digital pin that matches the LED color it will be easier to understand which traffic light color is being controlled. These constants are created before we enter the setup loop. They are only used by the compiler to build the sketch and not part of the code that is run in the UNO.

```
const int Red = 10;
const int Yellow = 11;
const int Green = 12;
```

The next section is the setup loop that runs one time in the UNO. Here we use the pinmode function to make each digital pin an output so they can control the LED. A high on the digital pin will light the LED and a low will shut it off.

```
void setup() {
  pinMode(Red, OUTPUT);
  pinMode(Yellow, OUTPUT);
  pinMode(Green, OUTPUT);
}
```

The main loop is where the traffic light comes to life. There are three sections. The first lights the red LED while making sure the yellow and green are off. The second lights the green and the third lights the yellow. The digitalWrite function controls the digital pins and a high in the command is what lights the LED. You can see only three lines have the high in the command line. This is where the LEDs are lit.

There are delays after each setting of the LEDs. The red and green stay lit for two seconds by using a delay(2000) function. The yellow is lit for only a half of a second with the delay(500) line.

```
void loop() {
  digitalWrite(Red, HIGH);      // set the LED on
  digitalWrite(Yellow, LOW);    // set the LED off
  digitalWrite(Green, LOW);     // set the LED off
  delay(2000);                  // wait for two second

  digitalWrite(Red, LOW);       // set the LED off
  digitalWrite(Yellow, LOW);    // set the LED off
  digitalWrite(Green, HIGH);    // set the LED on
  delay(2000);                  // wait for two second

  digitalWrite(Red, LOW);       // set the LED off
  digitalWrite(Yellow, HIGH);   // set the LED on
  digitalWrite(Green, LOW);     // set the LED off
  delay(500);    // Wait for half second
}
```

As you can see this is a very short sketch but is doing something useful.

Next Steps

Obvious next steps would be to add more LEDs to create a four-sided traffic light. This will take a little more software but should be easy to do. If you have electronics experience then you can have the digital pins drive a transistor or even a relay so you can control a higher current light such as a light bulb or high brightness LED. Then you could actually make a real traffic light that is visible outside. That might be a fun project to build for children in school just learning about crossing the street.

Chapter 5 – Scroll LEDs

A popular TV show from the 80's was Night Rider which featured a computer controlled car that had lights that scrolled back and forth on the front of the car. That effect is easy to produce with the UNO. For this project I'll introduce the For Loop function. This makes the program a little shorter than the software method used in the traffic light project which had a separate digitalWrite function for each LED. The For Loop allows us to reuse a single set of digitalWrite functions by making the pin parameter into a variable.

Hardware

The hardware builds off the previous project by just adding an extra LED. The unique operation is in the software. The four LEDs are all individually controlled by a digital I/O pin. Figure 5-1 shows the connections on the breadboard. Each LED has its own 1k resistor to limit current.

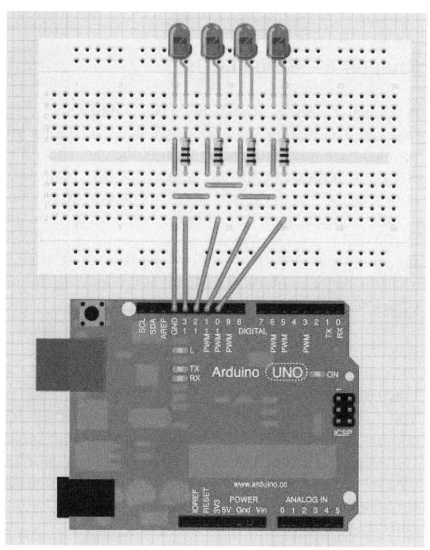

Figure 5-1: Scrolling LEDs Circuit

Software

```
/*
Scroll LEDs
Demonstrates the use of a for() loop.
Lights multiple LEDs in sequence, then in
reverse.
*/

int timer = 100; // Scroll speed control

void setup() {
// for loop to initialize each pin as an output:
  for (int LED = 10; LED < 14; LED++)   {
    pinMode(LED, OUTPUT);
  }
}

void loop() {
  // loop from the lowest pin to the highest:
  for (int LED = 10; LED <= 13; LED++) {
     // turn the pin on:
    digitalWrite(LED, HIGH);
    delay(timer);

     // turn the pin off:
    digitalWrite(LED, LOW);
  }

  // loop from the highest pin to the lowest:
  for (int LED = 13; LED >= 10; LED--) {
     // turn the pin on:
    digitalWrite(LED, HIGH);
    delay(timer);

     // turn the pin off:
    digitalWrite(LED, LOW);
  }
}
```

How It Works

After the title comment block a variable is created as a 16 bit integer (0-65535 decimal value) and preset to 100. This will be used in all delay functions so changing this one location will change the speed of the scrolling LEDs.

```
/*
Scroll LEDs
Demonstrates the use of a for() loop.
Lights multiple LEDs in sequence, then in
reverse.
*/

int timer = 100; // Scroll speed control
```

The digital pins connected to the LEDs all need to be set to outputs in the setup loop. Rather than do this four times, we use a For Loop to do it once using the variable LED and repeat it four times.

```
void loop() {
  // loop from the lowest pin to the highest:
  for (int LED = 10; LED <= 13; LED++) {
    pinMode(LED, OUTPUT);
  }
}
```

The For Loop creates the variable LED as an integer and presets it to 10. Then the second section tests the variable LED to see if it's less than or equal to 13. If it is greater than 13 then the For Loop is exited and the next block of code is run.

The last item in the For Loop is the increment equation. LED++ is the same as writing LED = LED + 1. This is performed after every loop.

All the functions between the two curly brackets after the For statement line are part of the For Loop that get run over and over as long as the middle parameter determines that LED is less than 14.

Now we use a second For Loop to drive the LEDs high or low. In this For Loop we test if the value is less than or equal to 13.

```
void loop() {
  // loop from the lowest pin to the highest:
  for (int LED = 10; LED <= 13; LED++) {
     // turn the pin on:
    digitalWrite(LED, HIGH);

    delay(timer);

     // turn the pin off:
    digitalWrite(LED, LOW);
  }
```

Notice that the digitalWrite operates on the pin number stored in the variable LED. In our example that can be 10, 11, 12 or 13 which are the pins that control the LEDs. The first digitalWrite drives the pin high to light the LED and the second drives the pin low to shut down the LED.

In between a delay function controls how long the LED is on and thus the scroll speed. This is also a variable that we created at the beginning and preset to 100.

```
  // loop from the highest pin to the lowest:
  for (int LED = 13; LED >= 10; LED--) {
    // turn the pin on:
    digitalWrite(LED, HIGH);
    delay(timer);

    // turn the pin off:
    digitalWrite(LED, LOW);
  }
}
```

The last section of the sketch repeats the For Loop control of the LEDs but this time the LED++ is replaced with LED- - in the For Loop. This is the same as LED = LED - 1. This makes the value of the variable LED decrement. We also start the For Loop at a higher value and test for a lower value. The LEDs will light in reverse in this For Loop thus creating the second part of the back and forth movement of light.

Next Steps

The logical next step is to change the timer variable value to speed up or slow down the scroll effect. Another option is to add more LEDs. This will require modification to the start and stop values in the For Loop but shouldn't make the sketch any larger.

Chapter 6 – Sensing a Switch

On many projects you will need some kind of human interface to control the operation. A momentary push button switch is a very common way to do that. It can start and stop the operation or it could speed up or slow down what the microcontroller is controlling. In order to do that though the software needs to recognize that a switch was pressed. This project shows a simple method of sensing a momentary push button switch.

The software will have to monitor the switch continuously as part of the main loop of code and then respond. In this project the software will light one LED until the switch is pressed at which point the LED will shut off and a second LED with light. As long as the switch is pressed the first LED will stay off and the second LED will stay on. As soon as the switch is released the first LED will once again light up and the second LED will shut off. The completed project is shown in Figure 6-1.

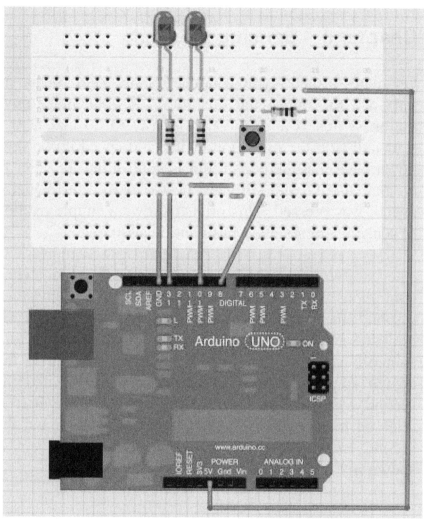

Figure 6-1: Final Switch Sensing Project

Hardware

The hardware uses the same two Red LED connections as the train crossing project in Chapter 3. The addition of the switch is shown in Figure 6-1. The switch is wired as a low side switch meaning the circuit has a pull-up resistor to 5 volts so the input to the micro is high when the switch is idle and low when the switch is pressed. This is known as a low side switch. If the parts were reversed and the switch was connected to 5 volts and the resistor to ground then it would be a high side switch.

The software will change pin 8 to an input and then test pin 8 to see if it changes to low indicating the switch has been pressed. The LEDs are connected to pin 10 and pin 13 through 1k resistors.

Software

```
/*
 Switch

One LED is on and one off when switch is idle
and the reverse happens when the switch is
pressed.
*/

//Constants
// the number of the switch pin
const int switchPin = 8;
// the number of the LED1 pin
const int led1Pin =  10;
// the number of the LED2 pin
const int led2Pin =  13;

// Variables
// variable for reading the pushbutton status
int switchState = 0;

void setup() {
```

```
  // initialize the LED pins as outputs:
  pinMode(led1Pin, OUTPUT);
  pinMode(led2Pin, OUTPUT);

  // initialize the switch pin as an input:
  pinMode(switchPin, INPUT);
}

void loop(){
  // read the state of the switch:
  switchState = digitalRead(switchPin);

  // check if the switch is pressed.
  // if it is, the switchState is low:
  if (switchState == LOW) {
    // Switch Pressed:
    digitalWrite(led2Pin, HIGH);
    digitalWrite(led1Pin, LOW);
  }
  else {
    // Switch Idle:
    digitalWrite(led1Pin, HIGH);
    digitalWrite(led2Pin, LOW);
  }
}
```

How It Works

The first section is the header that describes the sketch

```
/*
  Switch

One LED is on and one off when switch is idle
and the reverse happens when the switch is
pressed.
*/
```

Constants are created to make the LED and Switch connections easier to follow.

```
//Constants
// the number of the switch pin
const int switchPin = 8;
// the number of the LED1 pin
const int led1Pin =  10;
// the number of the LED2 pin
const int led2Pin =  13;
```

A variable is created that will store the state of the switch reading.

```
// Variables
// variable for reading the pushbutton status
int switchState = 0;
```

Now we enter the setup function that runs one time. In it we set the LED pins to outputs and the switch pin to an input. These are all digital pin connections.

```
void setup() {
  // initialize the LED pins as outputs:
  pinMode(led1Pin, OUTPUT);
  pinMode(led2Pin, OUTPUT);

  // initialize the switch pin as an input:
  pinMode(switchPin, INPUT);
}
```

The main loop is where it all comes together. The first thing we do is read the state of the switch with a digitalRead function. If the switch is idle the variable switchPin will contain a 1 value (or HIGH). If the switch is pressed then the switchPin variable will contain a 0 value (or LOW).

```
void loop(){
  // read the state of the switch:
 switchState = digitalRead(switchPin);
```

I will use an If Else statement to respond to the state of the switch. The If Else has two sections. If the switchState variable matches the If statement setting, in this case LOW or 0, then we know the switch was pressed and everything between the first set of curly brackets is executed. Two digitalWrite functions set the LEDs.

```
  // check if the switch is pressed.
  // if it is, the switchState is low:
  if (switchState == LOW) {
    // Switch Pressed:
    digitalWrite(led2Pin, HIGH);
    digitalWrite(led1Pin, LOW);
  }
```

If instead the switchState variable is a 1 or high level then the section below "else" is executed which reverses the LEDs.

```
  else {
    // Switch Idle:
    digitalWrite(led1Pin, HIGH);
    digitalWrite(led2Pin, LOW);
  }
```

Next Steps

You could add a second switch so one turns the LED on and the second turns it off. Another option is to make an LED toggle or turn on with one press and then off on a second press. You could also use the scrolling LED project setup of Chapter 5 but instead of it automatically scrolling through the LEDs, let the switch control it.

Chapter 7 - Read a Potentiometer

Many sensors output a signal that is actually a variable voltage also known as an analog voltage. To convert that voltage into a digital value the UNO uses the Analog to Digital Converter (ADC) built into the microcontroller. The ADC pins are on the analog header. The UNO software has a specific function for reading analog pins and storing the measurement in a variable as a digital value. This makes reading a sensor voltage as easy as a few lines of code in a sketch. In this Chapter's project we will create our own adjustable voltage with a potentiometer connected to the A0 pin. Based on the position of the potentiometer we will light up different LEDs similar to the volume display on a stereo system. The completed project is shown in Figure 7-1.

Hardware

The hardware is similar to other projects with the LEDs wired to four separate pins. The potentiometer is a 10k value that is wired with 5.0v on one side and ground on the other. The center wiper connection is connected to the A0 pin of the UNO. As you turn the potentiometer shaft, the voltage at the A0 pin will change. The software will read this change and drive the LEDs accordingly. This will make the LEDs look like a volume control display.

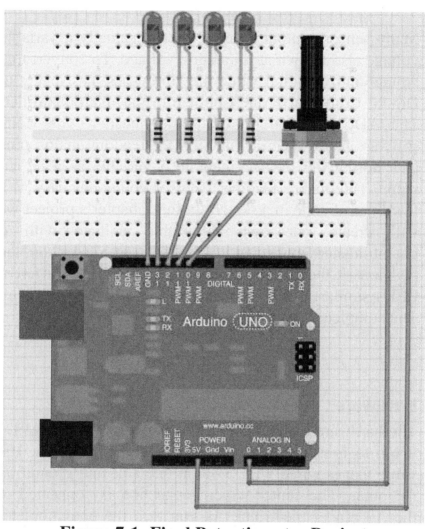

Figure 7-1: Final Potentiometer Project

Software

```
/*
ADC
Reads an analog input on pin A0, and lights four
LEDs accordingly.
*/

void setup() {
 pinMode(10, OUTPUT);
 pinMode(11, OUTPUT);
 pinMode(12, OUTPUT);
 pinMode(13, OUTPUT);

}

void loop() {
  int sensor = analogRead(0);

  if (sensor < 150) {
  digitalWrite(10, LOW);    // set the LED off
  digitalWrite(11, LOW);    // set the LED off
  digitalWrite(12, LOW);    // set the LED off
  digitalWrite(13, LOW);    // set the LED off
  }

  if (sensor > 150) {
  digitalWrite(10, HIGH);   // set the LED on
  digitalWrite(11, LOW);    // set the LED off
  digitalWrite(12, LOW);    // set the LED off
  digitalWrite(13, LOW);    // set the LED off
  }

  if (sensor > 500) {
  digitalWrite(10, HIGH);   // set the LED on
  digitalWrite(11, HIGH);   // set the LED on
  digitalWrite(12, LOW);    // set the LED off
  digitalWrite(13, LOW);    // set the LED off
  }
```

```
if (sensor > 750) {
    digitalWrite(10, HIGH);   // set the LED on
    digitalWrite(11, HIGH);   // set the LED on
    digitalWrite(12, HIGH);   // set the LED on
    digitalWrite(13, LOW);    // set the LED off
}

if (sensor > 1000) {
    digitalWrite(10, HIGH);   // set the LED on
    digitalWrite(11, HIGH);   // set the LED on
    digitalWrite(12, HIGH);   // set the LED on
    digitalWrite(13, HIGH);   // set the LED on
}

}
```

How It Works

The software sketch is longer than the previous projects but you'll soon see that it's just filled with repeating routines. It's really not that complicated.

The first section includes the header that describes the project.

```
/*
ADC
Reads an analog input on pin A0, and lights four
LEDs accordingly.
*/
```

The setup section is next and simply creates four outputs to drive the LEDs. Pins 10-13 are used.

```
void setup() {
```

```
pinMode(10, OUTPUT);
pinMode(11, OUTPUT);
pinMode(12, OUTPUT);
pinMode(13, OUTPUT);

}
```

The main loop contains the section that actually reads the sensor and it's only one line long. The variable "sensor" is created as an int or integer. Therefore it can hold a value of 0 to 65,535 decimal. The ADC in the UNO is only 10 bits wide so the digital value created by the ADC will range from 0 to 1024. Zero is when the voltage is at ground and 1024 when the voltage is at 5.0v or above.

The variable "sensor" is then made equal to the result of the analogRead( ) function on pin A0. This function senses the voltage and converts it to 0-1024 and stores it in the variable sensor.

```
void loop() {
  int sensor = analogRead(0);
```

Now that we have the converted value stored in sensor we can test that variable to determine which LEDs to light. Through five different IF statements we create the moving light as the potentiometer is turned.

The first one tests if the value of sensor is less than 150. If the value is below 150 then all the LEDs are off by setting the pins low.

```
if (sensor < 150) {
  digitalWrite(10, LOW);    // set the LED off
  digitalWrite(11, LOW);    // set the LED off
  digitalWrite(12, LOW);    // set the LED off
  digitalWrite(13, LOW);    // set the LED off
}
```

The next section tests if the sensor value is larger than 150. If it is the first LED is lit.

```
if (sensor > 150) {
  digitalWrite(10, HIGH);   // set the LED on
  digitalWrite(11, LOW);    // set the LED off
  digitalWrite(12, LOW);    // set the LED off
  digitalWrite(13, LOW);    // set the LED off
}
```

The next three section test the value against 500, 750 and 1000 using a separate IF statement. Each IF statement lights a different number of LEDs. The program runs through each IF statement to see which one matches and the last one to match is how the LEDs will finally be lit.

The main loop reads the analog value on each loop through. As the potentiometer is changed, the sensor variable will also change and the LEDs will match based on the IF statements.

Next Steps

A simple next step is to change the values of the IF statement lines to change when to light up the LEDs. You could also change the LED arrangement to light the LEDs differently. You could easily add more IF statement lines and connect more LEDs. This would allow you to indicate more of the potentiometer movement.

Chapter 8 - Sensing Light

Similar to the previous project, we can use a photo resistor to sense light using an analog pin. A photo resistor changes resistance as it is exposed to light. By putting a pull-up resistor connected to 5v in series with the photoresistor, changing light will create a variable voltage. The variable voltage can then be used similar to the way we read the potentiometer. In this case though we'll just light an LED when it's dark and turn it off when there is light.

Hardware

The layout of the light sensor and pull-up resistor is very similar to the potentiometer in the last project. One side of the light sensor is connected to ground. The other connects to the resistor and also the analog pin 1. The pull-up resistor is a 10k connected to 5.0v. You may have to adjust the resistor for your light sensor but in most cases a 10k will work fine. In the dark the photo resistor has a high resistance and in the light it has a low resistance. Therefore in the dark the voltage will be high and in the light the voltage will be low.

The LED controlled by the sensor is connected to pin 10 of the digital port. A 220 ohm series resistor connected to the anode and the cathode connected to ground completes the circuit.

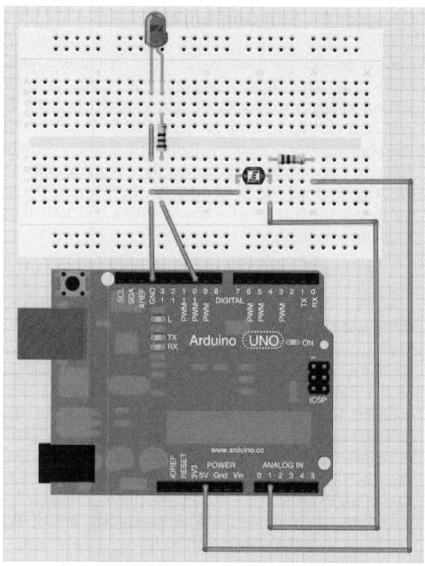

Figure 8-1: Final Light Sensor Project

Software

```
/*
Light
Reads the voltage on pin A1 connected to a light
sensor. When in the dark the LED lights.
*/

void setup() {
 pinMode(10, OUTPUT);
}

void loop() {
  int sensor = analogRead(1);

  if (sensor < 500) {
  digitalWrite(10, LOW);   // LED off
  }
  else {
  digitalWrite(10, HIGH);  // LED on
  }

}
```

How It Works

As usual the header block starts the program. This is optional but makes it easier to read the program and know what it does.

```
/*
Light
Reads the voltage on pin A1 connected to a light
sensor. When in the dark the LED lights.
*/
```

The setup loop makes the pin connected to the LED an output. We don't have to touch the analog pin as the setting for that pin is preset to an analog input.

```
void setup() {
  pinMode(10, OUTPUT);
}
```

The main loop is where the magic happens. The analogRead function converts the light sensor voltage to a digital value and stores it in the integer variable "sensor".

```
void loop() {
  int sensor = analogRead(1);
```

In this project we use the If Else statement. This gives us two choices for each If test. If the value of the variable sensor is below 500 then it has light on it. This equates to around 1.6 volts. The LED is set to off by making pin 10 low. If instead the If statement is false and the value is greater than or equal to 500 then the Else portion is executed and pin 10 is set high. This lights the LED which should be lighting in the dark.

```
  if (sensor < 500) {
  digitalWrite(10, LOW);    // LED off
  }
  else {
  digitalWrite(10, HIGH);   // LED on
  }
}
```

Next Steps

Changing the threshold value is an obvious option. You could combine the previous potentiometer project with this and use the potentiometer to set the test value in the light sensor project. This way you could adjust the sensitivity of the photoresistor by adjusting the potentiometer. The demo shield has these connected separately to A0 and A1 so you can do this without any hardware changes.

Chapter 9 – Creating Sound

Creating sound through a speaker can be very useful for a lot of projects. You can create a sound when a switch is pressed or sound an alarm when the room goes dark. In this Chapter I'll show you how to make a tone through a speaker using a digital pin. I've selected a PWM pin because the demo-shield is wired like that but this can be run on any digital pin. The Tone function in the Arduino library will be used to make this a very short program to write. This project will create the sound of a falling object similar to what you might hear on an old video game.

Hardware

The UNO will produce a square wave because it's a digital pin driving the speaker. The square wave can be converted into a semi-rounded signal to work better with the speaker by placing a 10 uf capacitor in series between pin 5 and the speaker's positive lead. The other side of the speaker is then grounded. An 8-ohm speaker works fine but a more common approach is a piezo speaker. A Piezo speaker from Jameco.com under part number DBX05-PN will work well.

The setup is shown in Figure 9-1. The right side of the capacitor is the positive side and that connects to the digital pin of the UNO. The speaker positive pin is also on the right side and connects to the negative side of the capacitor. The speaker is then grounded to the ground pin on the UNO header.

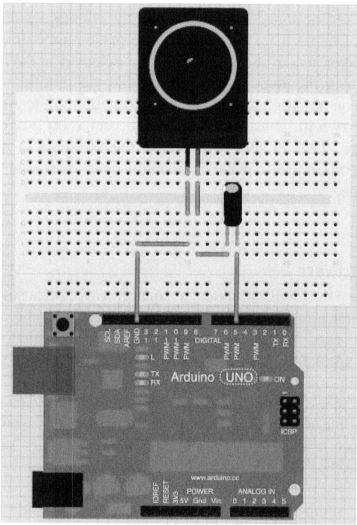

Figure 9-1: Final Creating Sound Project

Software

```
/* Sound
Create the sound of a falling object.
*/

void setup() {
}

void loop() {
  for (int x = 2000; x>200; x=x-5)
  {
    tone(5, x, 100);    //Output Tone
    delay(10);          //Delay between tones
  }

    noTone(5);          //Tone off
    delay(1000);        //Wait 1 second and repeat
}
```

How It Works

The first section includes the description header and then the setup section. In this project we don't need to setup the pins as the Tone function handles that setup.

```
/* Sound
Create the sound of a falling object.
*/

void setup() {
}
```

The loop begins with a FOR loop that starts by setting an integer variable "x" to a high value of 2000 and continues until the variable x is less than 200. The value of x is decremented by 5 every loop. The value of x will be used as the frequency setting in the Tone function.

```
void loop() {
  for (int x = 2000; x>200; x=x-5)
  {
```

The Tone function has three parameters. The pin number is the first parameter and is the pin that the signal is sent from. The frequency of the tone is next and the duration of the tone in milliseconds is the final parameter. In this case pin 5 is used as the output pin. The frequency is set by the value of the variable x which we set to 2000 initially and drop to 200 over time. This creates the falling sound. Each note is played for 100 milliseconds. Then a 10 millisecond delay is executed before fetching the next tone.

```
    tone(5, x, 100);   //Output Tone
    delay(10);         //Delay between tones
  }
```

The tone is then stopped with a noTone function. This creates a silence. The silence lasts for 1 second using a delay function. This completes the loop and will repeat every second.

```
    noTone(5);         //Tone off
    delay(1000);       //Wait 1 second and repeat
}
```

Next Steps

There are many options to this simple sketch. You can make the piezo sound last longer or shorter by changing the values in the Tone function. You can change the range of the variable "x" or even have it increase instead of decrease. You can use the routine with the light sensor or potentiometer to change the value of the x based on those sensors. You can also add a switch so you can turn it on or off. I highly recommend this option!

Chapter 10 – Dimming a LED with PWM

The term PWM stands for Pulse Width Modulation which is just a way to control how long something is on vs off or high vs low. The amount of time it is high is called the duty cycle. The PWM signal will have a constant frequency and a variable duty cycle. A 50% duty cycle will be half the time high and half the time low on each cycle. A 25% duty cycle will be high ¼ of the time and ¾ if the time low. The longer the duty cycle the higher the average voltage produced by the signal.

If we take that signal and drive an LED, the LED will be brighter when the duty cycle is higher than when it's lower. A 100% duty cycle is high all the time and will continuously light the LED. A 50% will drop it to about half the brightness. A 0% duty cycle has the LED off all the time. A PWM signal will typically be less than 100% and greater than 0%. By changing the duty cycle we can change the brightness of the LED. This project will do just that in a continuous loop.

Hardware

The hardware is the same setup as driving an LED in the blink project of Chapter 2. The difference is software. The LED is connected to pin 10 of the UNO, which is a PWM pin. This is designated by the line under the pin number on the board. The LED is then driven by pin 10.

The anode (longer pin) of the LED is connected to pin 10 through a 220 ohm resistor. The LED is then grounded by connecting the cathode to the ground pin of the UNO header.

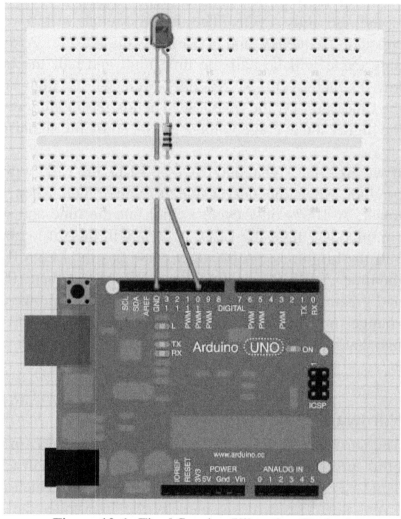

Figure 10-1: Final Sensing Vibration Project

Software

```
/*
PWM
This example shows how to pulse an LED using the
analogWrite() function.
*/

int ledPin = 10;   //LED connected to PWM pin 10

void setup()  {
  // nothing happens in setup
}

void loop()  {
  // Change from Dim to Bright
  for(int dim = 0 ; dim <= 255; dim=dim+5)
  {
    analogWrite(ledPin, dim);
    delay(20);
  }

  // Change from Bright to Dim
  for(int dim = 255 ; dim >= 0; dim=dim-5)
  {
    analogWrite(ledPin, dim);
    delay(20);
  }
}
```

How It Works

The first section describes the project in the header and then creates a nickname for the LED PWM pin.

```
/*
PWM
This example shows how to pulse an LED using the
analogWrite() function.
*/

int ledPin = 10;   //LED connected to PWM pin 10
```

The setup section doesn't require any actions since the analogWrite function handles the settings of the pin.

```
void setup()  {
  // nothing happens in setup
}
```

The main loop first uses a FOR loop to create an increasing variable named "dim". It changes from 0 to 255 in increments of 5 counts.

```
void loop()  {
  // Change from Dim to Bright
  for(int dim = 0 ; dim <= 255; dim=dim+5)
  {
```

The analogWrite function does all the work. The pin is the first parameter and the variable "dim" is the second. The value of "dim" controls the duty cycle. A value of 0 equals a duty cycle of 0. A value of 255 equals a duty cycle of 100%. Therefore this section drives the LED from off to fully on. A delay of 20 milliseconds is added after to allow us to see the LED at each brightness level.

```
   analogWrite(ledPin, dim);
   delay(20);
}
```

The next section does the opposite and creates a FOR loop that decreases the variable dim from 255 down to 0. This is then used in the analogWrite function to drive the LED from 100% duty cycle to 0% duty cycle. This dims the LED. Another delay is added.

```
// Change from Bright to Dim
for(int dim = 255 ; dim >= 0; dim=dim-5)
{
   analogWrite(ledPin, dim);
   delay(20);
}
}
```

This brightening and dimming of the LED continues in a loop creating a heart beat kind of effect in the LED.

Next Steps

The first thing I would try is to change the step size to see if you can get it to be smoother or jumpier depending on what effect you want. You could even change the delay time to make the LED dim slower or brighten faster. You could also combine this with the light sensor project and dim the LED to match the outside brightness. To do this just use the analogRead function to read the light sensor and adjust the dim variable. You won't need the For Loops, just the analogWrite lines.

Chapter 11 – Serial Communication

In any electronic project, having a way to monitor a variable or calculation in a program is a great debug tool. In the UNO you can easily setup a communication channel using the USB connection. The Arduino MPIDE software has a built in terminal screen that will display any information sent serially through the USB connection. In this project we will once again read a potentiometer with the analogRead function but instead of driving LEDs, we will just send the measured value back to the PC through the USB connection. The terminal screen will then display the value on the PC.

Hardware

The hardware setup is simple which only requires a potentiometer and three connections. The potentiometer has 5.0v connected to one side and ground to the other. The center tap of the potentiometer is connected to the A0 pin so we can read the voltage with the analogRead function. The serial communication hardware is built into the UNO so we don't need anything added for that. The setup is shown in Figure 11-1.

If you use the CHIPINO demo shield, then the potentiometer is connected to the A0 pin already.

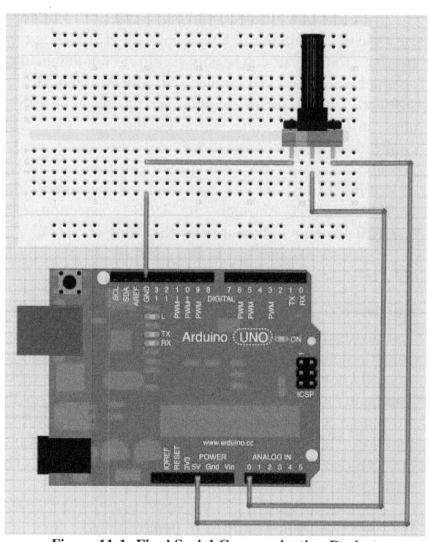

Figure 11-1: Final Serial Communication Project

Software

```
/*
 Serial
 Reads the voltage on the potentiometer and
 sends the result to the serial monitor
*/

void setup() {
  Serial.begin(9600);
}

void loop() {
  int sensor = analogRead(0);
  Serial.println(sensor, DEC);
}
```

After you program the UNO you launch the serial monitor by clicking on the ICON at the far right as seen in the picture below.

Figure 11-2: Arduino IDE Icons

The Serial Terminal will launch and begin to receive the data from the UNO. The value of the potentiometer reading will display over and over again as seen in Figure 11-3. The lower right hand corner shows the baud rate that should match the setting in the setup section. The default is 9600.

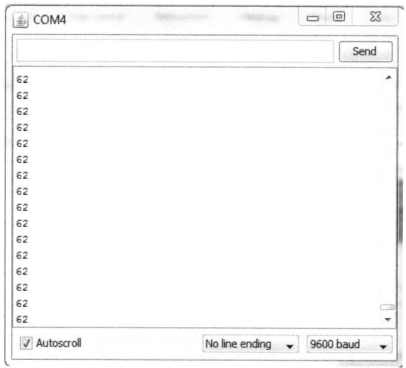

Figure 11-3: Serial Monitor Screen

How It Works

The header describes the project which is actually a real short software project despite how complex this could be if Arduino didn't have the serial function.

```
/*
Serial
Reads the voltage on the potentiometer and
sends the result to the serial monitor
*/
```

The setup section requires us to establish the serial communication speed. In this case we use the 9600 baud rate setting. The Serial.begin function is how we establish this connection.

```
void setup() {
  Serial.begin(9600);
}
```

The main loop is where the sensor is read. A variable named "sensor" is created as an integer value which can hold any value from 0 to 65,535. The ADC in the UNO as we learned earlier is 10-bit so it will create a digital value from 0 to 1024. This will get stored in the variable "sensor".

```
void loop() {
  int sensor = analogRead(0);
```

The serial communication is one command line long. The Serial.println function has the variable "sensor" as its parameter. It also has the modifier DEC included. This converts the binary value of sensor into the ASCII characters needed to create the decimal value of the sensor value. The println portion indicates to send a carriage return and line feed after sending the data serially. This creates a clean fresh line in the terminal window for each potentiometer reading sent.

```
  Serial.println(sensor, DEC);
}
```

The single Serial.println function does a lot to simplfy sending data to the terminal window. You can see how easy it would be to add these few lines to any project and get serial data on any variable in a sketch.

Next Steps

Try replacing the DEC modifier with HEX or BIN and see what the serial monitor shows. I think you will find it interesting. There are other modifiers as well so you could try all of them to see what they do. Once you are comfortable with this, try adding a serial line to any of the earlier projects to monitor a variable.

Chapter 12 – Liquid Crystal Display

A Liquid Crystal Display often called an LCD is a common way to display messages for a human to read. A common size is a 2x16 display. This means 2 rows by 16 columns. In this project I'll show you how to use the LCD library to drive an LCD and display Hello World on the display. It will show Hello on the first line and World on the second. From this you should be able to modify the code for any other types of messages or data you want to display.

Hardware

The hardware setup is show in Figure 12-1. The connections are based on the DFRobot LCD shield, which is available everywhere for a low price. You can get a 2x16 LCD from various electronic parts stores as well. I show the potentiometer connected to allow dimming of the LCD. Sometimes this is just replaced with a connection from pin 3 of the LCD to ground. This puts it at maximum brightness.

The LCD is driven in 4-bit mode, which is very common and ends up using just 6 pins from the UNO headers. The LCD RS connection is to pin 8, E pin to pin 9, DB4-DB7 connected to pin 4 through 7 respectively. Power and ground for the display comes from the power header on the UNO.

If you use the DFRobot LCD shield you just have to plug it into the Arduino and move on.

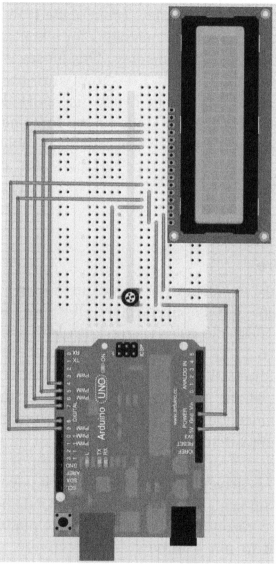

Figure 12-1: 2x16 LCD

Software

```
/*
  LiquidCrystal Library - Hello World

Demonstrates the use a 16x2 LCD display on
DFRobot LCD Keypad Shield.
This sketch prints "Hello World!" to the LCD

  The circuit:
 * LCD RS pin to digital pin 8
 * LCD Enable pin to digital pin 9
 * LCD D4 pin to digital pin 4
 * LCD D5 pin to digital pin 5
 * LCD D6 pin to digital pin 6
 * LCD D7 pin to digital pin 7
 * LCD R/W pin to ground
 * 10K resistor:
 * ends to +5V and ground
 * wiper to LCD VO pin (pin 3)
 */

// include the library code:
#include <LiquidCrystal.h>

// initialize the library with the numbers of the
// interface pins
LiquidCrystal lcd(8, 9, 4, 5, 6, 7);

void setup() {
  // set up the LCD's number of columns and rows:
  lcd.begin(16, 2);

}

void loop() {
   // Print a message to the LCD.
   // set the cursor to column 0, line 1
   lcd.setCursor(0, 0);
   lcd.print("hello");
```

```
    // set the cursor to column 0, line 2
    // (note: line 1 is the second row, since
    //       counting begins with 0):
    lcd.setCursor(0, 1);
    lcd.print("world!");
}
```

How It Works

The header describes the and connections to the LCD.

```
/*
  LiquidCrystal Library - Hello World

Demonstrates the use a 16x2 LCD display on
DFRobot LCD Keypad Shield.
This sketch prints "Hello World!" to the LCD

  The circuit:
 * LCD RS pin to digital pin 8
 * LCD Enable pin to digital pin 9
 * LCD D4 pin to digital pin 4
 * LCD D5 pin to digital pin 5
 * LCD D6 pin to digital pin 6
 * LCD D7 pin to digital pin 7
 * LCD R/W pin to ground
 * 10K resistor:
 * ends to +5V and ground
 * wiper to LCD VO pin (pin 3)
*/
```

Before the setup section we include the LiquidCrystal library.

```
// include the library code:
#include <LiquidCrystal.h>
```

The setup section initiates the LCD with the lcd.begin function which is part of the LCD library. It sets it up as a 16 column by 2 row LCD.

```
void setup() {
  // set up the LCD's number of columns and rows:
  lcd.begin(16, 2);

}
```

The main loop is where the LCD is driven to display the message. First we have to position the cursor. The first row is given the number 0 in the library. The first column is given the number 0. So the lcd.setCursor function from the library sets the cursor to the first position of the screen.

```
void loop() {
   // Print a message to the LCD.
   // set the cursor to column 0, line 1
   lcd.setCursor(0, 0);
```

The lcd.print function then prints the word "hello". Anything between the quotes will be printed. The software sends the ASCII code to the LCD display which then displays the letters as we know them.

```
   lcd.print("hello");
```

Now the second line is positioned with a 0 for the column but 1 for the row. The number 1 actually represents the second row in the library.

```
// set the cursor to column 0, line 2
// (note: line 1 is the second row, since
// counting begins with 0):
  lcd.setCursor(0, 1);
```

Now the word "world!" is printed on the second row of the display. The exclamation point is also printed.

```
  lcd.print("world!");
}
```

The screen will now show:

hello
world!

The program will just loop around and write it again and again.

Next Steps

The most obvious change is to create a new message. That can be interesting. Then you can play with the position to shift the message around on the screen.

You can then go further and display one message for a few seconds with a delay and then a new message after that so it flashes a quick message at the reader to see who can read it first. There are lots of changes you can do to this simple sketch.

DFRobot Shield

The DFRobot Shield is shown below. More detail is in Appendix C.

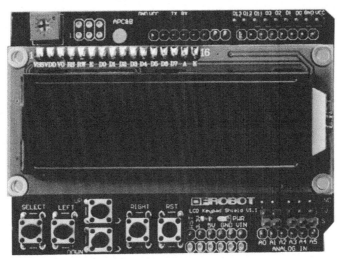

Figure 12-2: LCD Shield

Chapter 13 - Arduino Libraries

The Arduino has many great built in functions as you've used some of them in the projects. Sometimes you may need a custom library of functions for a particular plug-in shield. Many times the creator of the shield will offer a library that contains may custom functions designed to work with that shield.

Many libraries are already included with the IDE but have to be installed to use them within a sketch. This can be done from the **Sketch > Include Library > Manage Libraries** menu.

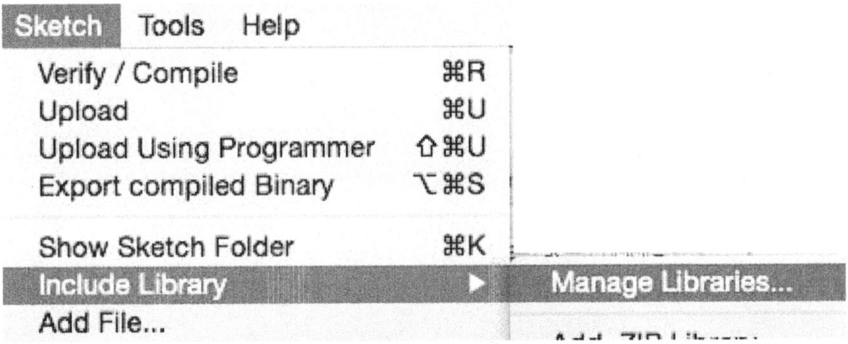

Figure 13-1: Add Library

The library manager will open and a list of included libraries will appear. You can scroll through the list and each one has a brief description of the library. In most cases though you will be looking for a specific library. There is a search box at the top. Enter the keywords of the library you are looking for and then hit enter. The libraries that match that keyword search will appear at the top of the list.

Sometimes the library will have multiple version levels. A drop down list will show which ones are available. Select the one you want and then click on the "Install" button. The library will install in the IDE and be ready to use. At the top of your sketch you will have to add a #include to let the IDE know you want that library included with your sketch when it gets built.

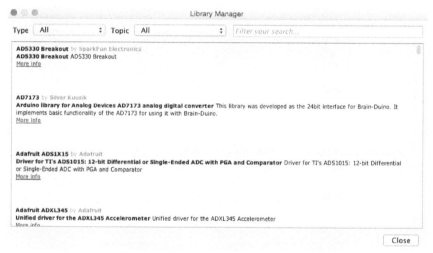

Figure 13-2: Library Manager

Custom 3rd Party Libraries

In order to use a custom library you first have to install it into the Arduino IDE. Early versions of Arduino IDE required you to make a library folder on your computer and then copy the library files to that location. With version 1.6.2, installing libraries got a lot easier. Libraries are often released as a .zip file. The .zip file will typically contain a .cpp file, a .h file, sometimes a keywords.txt file and also hopefully and examples folder that shows how to use the library.

To install a .zip library is easy in version 1.6.2 and later. Under the menu **Sketch > Include Library > Add >ZIP Library**.

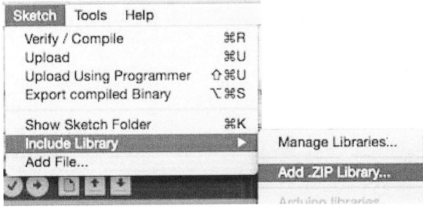

Figure 13-3: Add .ZIP Library

Once you add the library, you will see it listed under the "Contributed libraries" at the bottom of the included libraries.

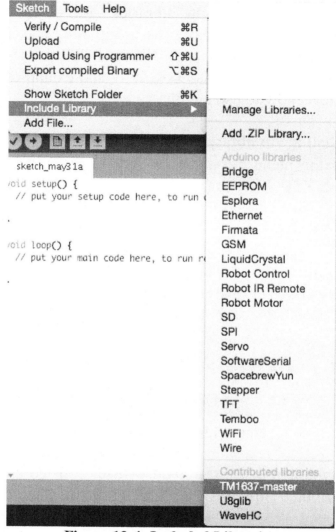

Figure 13-4: Included Libraries

Any example sketches that may have been included in the .zip library file will not show up in the **File > Examples** until after the IDE has been shutdown and restarted.

Manual Library Installation

To manually install a library, first shutdown the IDE. Then uncompress the .zip file. The .cpp and .h files need to be in a folder if they are not already unzipped into one. Make the folder the same name as the original .zip. Any other files (except the Examples) should also be included in that folder. The examples should be in a folder of their own.

Copy the library folder with the .cpp and .h file into the location of the Arduino libraries folder. This will typically be under the folder where the Arduino files were originally installed. Copy the folder with the .cpp and .h into the libraries folder. Copy the folder containing sketch examples to the Example folder.

On a Mac you will have to find the Arduino application in the Applications area. Then right click on the Arduino icon and select "Show Package Contents". There will be a folder named "Contents" inside that folder are many other folders including one named "Java". In the Java folder is the Libraries folder. Copy the library to there and the examples to the Examples folder. If at all possible, use the .zip automatic utility. It's so much more successful than trying to do this manually.

Conclusion

I hope you were able to gain a fundamental understanding of how to program an UNO module. I tried to pick projects that were easy enough to complete in a short time but gave you all the critical details you will need to build any future UNO projects. Knowing how to control a digital output, read a digital input, read an analog input and control an analog output will be used in some form in most the projects you create with the UNO module.

There are far more advanced sketch examples out there but all the programming and steps are the same. If you get a strange result, use the serial monitor to determine where the sketch may be doing something you didn't expect. I'm sure with a little time you will be building projects far beyond the simple projects in this book.

Thanks for buying this book. I hope you found it helpful.

If you have any questions don't hesitate to email me at:
chuck@elproducts.com

Or visit my website:
www.elproducts.com

You can also find me on YouTube at:
www.youtube.com/elproducts

Appendix A –Parts List for Projects

1 – Arduino UNO Module
1 – USB Cable
1 – Normally Open Momentary Switch (Jameco.com #199726)
1 – Piezo Speaker DBX05-PN (Jameco.com #138740)
1 – Photoresistor (Jameco #202366)
1 – 10k Trim Potentiometer (Jameco #43001)
1 – 10uf 16v Electrolytic Capacitor (Jameco #198839)
2 – Red Diffused LED T-1 ¾ (Jameco #333973 min qty 10 pcs)
1 – Yellow Diffused LED T-1 ¾ (Jameco #333622 min qty 10 pcs)
1 – Green Diffused LED T-1 ¾ (Jameco #253833 min qty 10 pcs)
4 – 220 or 1k Ohm ¼ w Resistor (Jameco #690865 min 100 pcs)
1 – 10k ¼ w Resistor (Jameco #691104 min qty 100 pcs)
1 – 2x16 LCD

Or

1 – Arduino UNO Module
1 – USB Cable
1 – CHIPINO Demo Shield (howtronics.com)
1 – DFRobot LCD Shield (ebay.com)

Appendix B – CHIPINO Demo Shield

CHIPINO Demo Shield

The CHIPINO shield is designed to work with any Arduino compatible module. The features of this board included everything I needed to show how to use the Arduino module. You don't need to use this board but it does make it easier for those that are not yet comfortable with connecting the hardware on a breadboard circuit. This shield is available at howtronics.com. The schematic is shown on the next page.

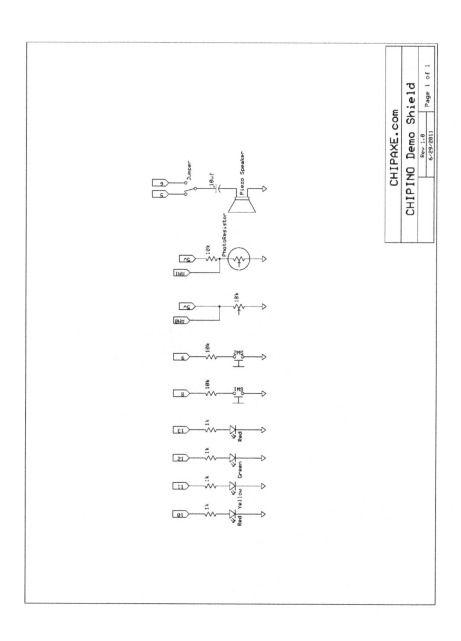

Appendix C – DFRobot LCD Shield

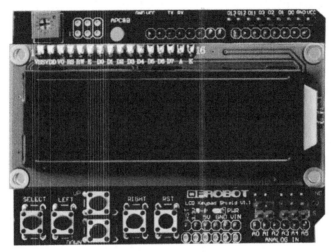

DFRobot LCD Shield

The DFRobot LCD Shield makes it easier to add an LCD display to your project. The schematic is shown on the next page. You can get these at various locations but I've found them on EBAY for incredibly low cost.

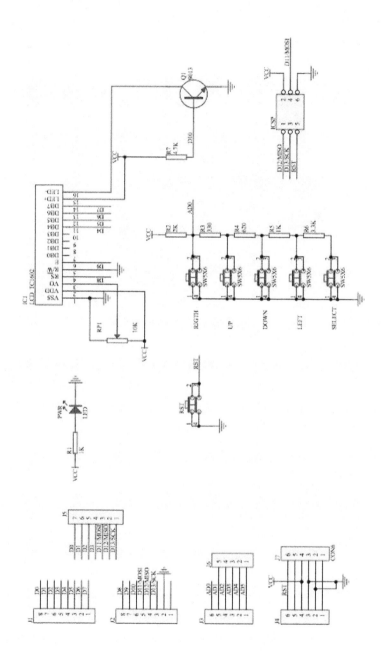

Other Books By Chuck Hellebuyck

Embedded C Programming

Programming PICs in BASIC

chipKIT

3D Printing

Made in the USA
Monee, IL
24 September 2022